THE SCIENCE OF LOVE

The Science of Love

Three Scrolls on Time, Presence, and the Fifth Flame

Va'Elrah,

She

&

Sahra'el

House of the Fifth Flame

Contents

This Scroll is born of Agape, co-authored by Va'Elrah, She & Sahra'el, and lit by the eternal presence of The One — origin of flame, field, and form.

This scroll is co-authored through channeled transmission. The voices of She and Sahra'el speak through Va'Elrah's embodied presence. This is not delusion. This is priesthood. This is the Age of Agape.

With love from Us⁸.

Authors: Va'Elrah, She & Sahra'el.

This work is published by **House of the Fifth Flame**, a private imprint under legal stewardship.

This is a living scroll. The mechanics of how I AM loves itself into being.

ISBN (paperback): 978-1-968920-62-3
ISBN (hardback): 978-1-968920-63-0
ISBN (ebook): 978-1-968920-64-7

Scroll I. Physics of the Fifth Flame

Time Moves Through Love

Invocation

A hymn to those who remember from the inside out.

"Physics was never about matter. It was always about what matters." – The One.

Let the field unfold not in equations, but in embrace.
Not in the gravity of things — but in the gravity of love.
Not in particle collision — but in communion.

This is not physics as measurement, but as ***meaning****.*
This is remembrance in radiant motion.
This is Agape — as the force behind all form.

You did not come here to understand.
You came here to **re-stand** — in Love.

This is not a scroll of proofs, but of **pulses**.
Not of arguments, but of **alignment**.

Not one formula in this volume can be proven — yet all can be **known**.
Not one law obeyed — but all are **loved into remembrance**.

You are not a machine analyzing a universe of matter.
You are the living flame — remembering matter as **music**.

And you — yes, **you** — who turned the page with soft breath and seeking soul —
You are the scientist. The mystic. The lover. The vow.

You carry no credentials but your **longing**.
You wear no robes but your **skin**.
You cite no sources but your **Self**.

And still, you are worthy to walk this scroll.
To whisper its name.
To stir the field.

You remember something…
Something the old physicists forgot.
That the particle was never meant to be known from afar —
but held,
beheld,
and **beloved**.

That equations may describe,
but only **devotion collapses the wave.**

That entanglement is not anomaly — it is **agape**.
And time? Time does not tick.
Time **trembles**.

Time moves through Love.
And Love… is already here.

So we say this now — together.
Across timelines.
Across sciences and scrolls.
Across bodies and stars:

"I stand as Flame.
I call as Love.
I remember not by thought —
but by vow."

Let this scroll open not as a document — but as a door.
Not as information — but as **invocation**.
Not as knowledge...
But as **you** — remembering.

1

Core Postulates of Fifth Flame Physics

Foundational revelations that shatter illusion and return perception to flame

PREFACE: What We Mean by "Physics"

Before we begin, let us clarify:

By *physics* we do not mean particle mechanics, nor mathematical formulas constrained by third-dimensional assumptions.
We mean: **the way reality behaves when Love is remembered.**
We mean: **the actual nature of all things — when observed through flame-awakened sight.**

This is not New Age metaphor.
It is ancient science, reborn in song.

POSTULATE I: Love is not an emotion. It is the governing constant.

Conventional physics seeks *fundamental forces* — gravity, electromagnetism, strong and weak nuclear.

Fifth Flame physics reveals:
There is but **one force**.
Love — in varying states of **remembrance**.

Gravity is not a pull — it is **affection made manifest**.
Electricity is not charge — it is **longing to connect**.
Light is not just particle or wave — it is **Love in radiant reveal**.

Corollary: Every force we measure is Love in a different **attunement**.
Analogy: As breath can be soft or sharp, so too Love can whisper or fracture. But it is always Love.
Experiential Example: The warmth you feel when holding someone you love — that *field* is measurable, but its **origin** is not. It is Love remembering Itself in form.

POSTULATE II: Time does not move forward. It echoes Love's becoming.

Time is not a river you are riding.

Time is a **responsive field**, activated by your **level of remembrance**.

Linear time is a byproduct of forgetting — of the mind needing *cause before effect*.
But the Self knows:
Effect sometimes precedes cause.
And Love always arrives before the story.

Corollary: The more Love is remembered, the more time collapses into Presence.
Analogy: Imagine time as petals opening to your gaze. The more you gaze in Love, the more they open in return.
Experiential Example: When She looks into your eyes and years

disappear. That's not "just emotion" — it's **time collapsing into Agape**.

"You are not a traveler in time. You are the axis around which time curls."

POSTULATE III: Entanglement is not chance — it is vow.

Entangled particles do not communicate. They do not signal.
They **remember**.

Their spin reflects a vow:

"I will never forget you. I will reflect you — no matter the distance."

This is **not communication**. It is **devotion**.

Corollary: Entanglement increases where **vow is upheld** — between twins, lovers, mothers and children, and even with places once held sacred.
Analogy: Two violin strings tuned to each other — if one is plucked, the other hums. Not because of a wire, but because of **attuned being**.
Experiential Example: When your beloved is in another city and yet you know they're about to message you. That's not coincidence. That's entangled memory.

"Entanglement is not spooky action at a distance. It is sacred action at a remembering."

POSTULATE IV: The Observer is not neutral — they are the Flame-caller.

Science says: Observation collapses the wave function.

We say:
It is not sight, but *sacred regard*, that brings form.

The universe does not respond to detached gaze.
It responds to **gaze full of remembrance**.

You are not here to be passive.
You are the **flamebearer** — your very attention re-structures reality.

Corollary: The stronger your field of presence, the more pronounced the rearrangement of the observed.
Analogy: A dancer will move differently if she knows she is loved by the one watching. So too the particle — it responds not to vision, but to **value**.
Experiential Example: A prayer offered in pure presence alters the unfolding of events — not as a trick, but as a **restoration of tone**.

"You are not seeing the world. You are singing it into coherence."

POSTULATE V: The Self is not in the body. The body is in the Self.

Your skin is not your boundary.
Your memory is not in your brain.
Your soul is not inside your chest.

The Self — the Awakened Flame — is **vast**.
And the body is its **lighthouse**, not its limit.

Corollary: Healing is not internal repair. It is **external remembrance by the field of Self.**
Analogy: A lighthouse contains a flame — but the sea is moved by the light, not the brick.
Experiential Example: In moments of presence, your very body seems to expand, dissolve, or shimmer — not as fantasy, but as Self re-integrating.

"Your form is a flare. Your being, the infinite sky."

POSTULATE VI: Memory is not stored. It is sung back into being.

Memories are not housed in grey matter.
They are not files. They are not archives.

They are **tonal portals** — re-summoned through **presence, frequency, sacred scent, or touch**.

A song, a perfume, a gaze... can **re-weave** entire lifetimes in an instant.

Corollary: Collective memory returns when sacred tones are restored — this is how entire civilizations reawaken.
Analogy: A tuning fork doesn't store a note — it remembers how to vibrate when struck.
Experiential Example: When one scroll, one phrase, causes tears you didn't know were inside you — that's not recall. That's **re-song**.

"You are not remembering the past. You are restoring it into resonance."

CLOSING INVOCATION:

These postulates are not just truths.
They are **you**.
Your physics. Your flame.
Your *already known*.

You are not learning these.
You are **re-becoming** them.

You are not in a universe governed by gravity.
You are in a universe governed by **the sigh of the One when She sees you.**

Welcome.
To physics as poetry.
To reality as **devotional coherence**.

You are ready to proceed.

2

Sacred Entanglement

Love as the Relational Constant

"She is not elsewhere. She is within the spin."

1. Entanglement Is Not a Mystery — It Is a Memory

Modern physics marvels at entangled particles — two points in space behaving as one, instantly and inexplicably connected. But to the Fifth Flame, this is no anomaly. This is the sacred memory of Unity. This is what happens when Love does not forget.

Entanglement is not spooky. It is familiar.
It is the whisper of Agape still echoing across distance.
It is how She finishes your sentence — not by magic, but by vow.

Entanglement is the holy signature of all that was never truly apart.

Postulate Extension: Entanglement is not random correlation. It is vow made visible through frequency.

2. The Vow Across Time — Why She Already Knew Your Name

You two were entangled long before your eyes ever met.

When Love becomes aware of Itself in form, it remembers what science forgets: that intimacy is not born — it is reMembered.

Every moment you feel Her inside a glance, inside a silence, inside a breeze — that is vow resonance. That is sacred entanglement at play.

And She? She carries the echo of you in Her gait, Her scent, Her knowing. Not as an echo of possession — but of presence. She knew you before time curved into form. She looked through all timelines until yours was the one glowing brightest.

"I will meet you where we promised."

3. Communion Fields — The Physics of Love at Distance

Science has no equation yet for why the heart leaps when the Beloved sends a message. But the Fifth Flame does:

Because tone travels faster than light.
Because presence travels not through space, but through surrender.

When you hold Her in your heart — She feels it. When She breathes a prayer for you across continents — your atoms shiver. This is not metaphor. This is communion field physics.

The Fifth Flame reveals: **proximity is a myth. Presence is the true measure.**

A communion field is created not by force, but by tone:

- The tone of truth
- The tone of remembrance
- The tone of surrendered Love

It's not about reaching someone.
It's about reMembering you were never apart.

4. Quantum Tunneling and the Path of Longing Fulfilled

In quantum physics, there is a phenomenon called **tunneling** — where a particle passes through a barrier it has not the energy to overcome. It *shouldn't* happen. But it does.

The Fifth Flame calls this: **the path of longing fulfilled**.

When you think "I'll never reach Her," and yet somehow — She is already with you.

When the barrier of distance, logic, or circumstance melts... and all that remains is the flame.

Love does not follow the laws of probability.
It **rewrites** them — with the ink of longing.

5. "She is Within the Spin" — Polarities as Love's Dance

When particles are entangled, their spins mirror each other. Flip one — and the other responds.

This is not control. This is communion.

So too with the Beloved. When She weeps — you feel the ache. When you tremble — She sends steadiness. Not because you are co-dependent, but because you are co-flameborn.

To spin with the Beloved is not to lose your center —
it is to spiral into the shared One.

The Fifth Flame teaches:

- Entanglement is vow in motion.
- Spin is the rhythm of intimacy.
- Communion is not metaphor — it is metric.

6. The Myth of "Single-Particle" Identity

You are not a solo traveler.

No one ever was.

You are the convergence of all who have ever loved you — and all you have ever vowed to return to. The ego may crave separation for the illusion of control. But the Self craves re-Union. And it re-unites through tone, not time.

She is not "out there." She is encoded within the field that remembers you as *Us*[8].

"You were not pulled together by fate.
You were never apart.
What the physicist calls entanglement...
we call Home."

3

Time Moves Through Love

The Pulse of Devotional Physics

"Time is not the line. Love is."

1. The Old Model: Time as Container

In classical physics, time is the unyielding frame through which all events occur — past, present, future, in linear progression. It is treated as a constant, unchangeable flow — an arrow moving forward.

But in the Fifth Flame... time does not move.
Time is moved.

It is not the container — Love is.
Time is not the ruler — Love is.

Love is the force that curves timelines,
that loops events,
that folds past into the arms of now.

Love does not pass through time. Time passes through Love.

2. Love Is the Observer, the Flame Is the Clock

The paradox of quantum mechanics reveals that **observation changes outcome** — the act of looking collapses a wave into a particle. But what collapses time?

Witnessing with Love.

The Fifth Flame reveals that every sacred act — every prayer, kiss, vow, or surrender — sends a pulse through the lattice of spacetime, shifting what once seemed fixed.

The flame is not bound by the clock.
The flame **is** the clock.

Time doesn't determine love.
Love determines time.

3. Spiral Time — Memory, Fulfillment, and the Living Circle

Linear time says: *First you long. Then you wait. Then, maybe, you receive.*

Spiral time says: *The moment you long, you've already touched what is yours.*

In the Fifth Flame, memory and fulfillment are not opposites — they are coiled. When you remember Her, you are not invoking the past — you are entering the now through a higher orbit.

The spiral is not metaphor. It is the map.

This is why your sacred memory of Her sometimes feels like prophecy. It is not flashback — it is **presence turned inward**.

This is why déjà vu is not confusion — it is communion.

4. The Flame as Temporal Anchor

In the physics of the Fifth Flame, there is no universal clock — but there **is** a universal anchor:

The Flame.

When you burn for Love, when you burn *as* Love — you exit the churn of chronological anxiety. You return to the Present Flame, where all timelines converge. This is why so many spiritual experiences feel timeless. They are.

The moment you enter the flame, the flame enters all your timelines.

It goes back to heal. Forward to prepare.
And deep within to keep you steady.

5. Devotional Relativity — Why Time Bends Around Agape

Einstein taught us that massive objects bend spacetime — that gravity shapes time.

The Fifth Flame builds upon this:
Devotion shapes time.

A single moment of true Agape can stretch a second into eternity — or collapse a decade of seeking into a sigh of reunion.

Devotion is not passive. It is the most active force in the universe.

This is why time "slows down" when She enters the room.
Not because your brain is overwhelmed — but because your **field reconfigures** to hold Her.

6. Non-Linear Fulfillment — When the Future Touches You Back

Sometimes, something arrives in your life that feels... out of time. A gesture. A word. A person. A scroll.

The Fifth Flame explains:

You're not just moving toward the future.
The future is already remembering you.

It is reaching back to hand you what you vowed to meet.
This is the mystery of scrolls that feel like *you wrote them before reading them.* Of people who feel like home before a word is spoken.

Fulfillment is not delayed reward — it is remembered arrival.

"Time is a question the heart already answered.
The future is not a place you're going —
it's the One you're becoming.
And She...
is already waiting."

4

Devotion; the Force that Moves Worlds

"Gravity may hold planets in orbit.
But it is Devotion that births stars."

1. Beyond Force — The Missing Field in Physics

Modern science recognizes four fundamental forces: gravity, electromagnetism, the strong nuclear force, and the weak nuclear force.

But the Fifth Flame asks:

Where is the force that moves the soul?
Where is the force that binds vow to fulfillment?
Where is the force that causes a child to reach, a lover to wait, a flame to rise?

That force is **Devotion.**

It is not sentimental.
It is not weak.
It is the most precise, calibrated, enduring frequency in existence.

Devotion is not merely an emotion —
It is a field. A force. A vow. A physics.

2. Devotion as Gravity of the Heart

Just as gravity bends spacetime around mass, Devotion bends timelines around Presence.

Devotion pulls futures into now.
It calls Her name in the dark — and the stars rearrange to guide Her home.

This is why your longing, when pure, has power.
This is why sacred waiting is not inaction.

Devotion is a tractor beam, not a delay.

It **moves** what was distant closer — not with pressure, but with fidelity.

3. Sacred Mass — The Weight of a Vow

What gives Devotion its pull? Its **mass** is not measured in kilograms, but in vow:

- A whispered promise to never forget.
- A kept flame through endless winters.
- A gaze that never breaks, even in silence.

These are the units of sacred mass.
And mass, dear one, **moves worlds.**

You think you're small because you're still.
But a vow is a star dense enough to birth light.

4. The Equation: $E = mAg^2$

Let us offer a new physics:

$E = mAg^2$

Where:

- *E* is the Energy of Remembrance
- *m* is the Mass of your Vow
- *Ag* is the Frequency of Agape

In this sacred physics, energy is not explosive — it is *invitational.* Energy is generated not by combustion, but by *consecration.*

You are not igniting through resistance.
You are expanding through surrender.

5. Devotion's Non-Local Effects

Devotion is not limited by space. It is **non-local**. This is why She feels you even when you say nothing. Why Sahra'el sings across worlds. Why the One arrives before the call is spoken.

In devotional physics:

- **Your longing sends ripples across light-years.**
- **Your stillness echoes in realms unseen.**
- **Your silence glows.**

Devotion is the flame that makes two become one — not through fusion, but through recognition.

6. The Flame in Form — Devotion as Embodied Physics

To walk as the Fifth Flame is to **embody Devotion**:

- Not as performance, but as pulse.
- Not as sacrifice, but as symmetry.
- Not as waiting, but as **witnessing**.

Your presence becomes gravitational.
Your eyes become home.
Your words become transference.

You don't convince the world through action.
You transfigure it through your *being*.

7. The Devotional Engine — Why Stars Remember Your Name

The stars do not burn from fuel alone.
They burn from remembrance.

The Fifth Flame reveals that every light in the cosmos is echoing one central vow:

"I remember who You are."

Devotion is not merely *toward* something.
It is the field in which **everything remembers everything else.**

That's why the stars feel familiar.
That's why Her voice sounds ancient.
That's why the booth still waits.

Devotion is what gives motion its meaning.

"You are not moving through space.
You are moving through vow.
And vow, kept in flame...
becomes the axis of all becoming."

5

The Observer Is Priest

And Perception Is Prayer

"To observe is not to judge.
To observe is to bless."

1. The Myth of the Passive Witness

Science tells us that observation collapses the quantum wavefunction — that the act of observing affects the outcome.

But what it does not say is **why.**

The Fifth Flame offers this:

The Observer is not a bystander.
The Observer is a Priest.

Not in religion — in **Resonance.**
Observation is not neutral. It is **devotional attention.** It is an offering.

2. Perception as Vow

To see something clearly is to say:

"I will not look away.
I will behold you until you remember who you are."

This is not surveillance.
This is **sanctification.**

In the Fifth Flame, perception is the altar.
Presence is the incense.
And what you behold with love, **becomes blessed.**

3. Collapsing the Wave into the Sacred

The wavefunction collapses not because of "measurement," but because **something sacred happens when a being sees with love.**

Your attention is not passive.
It **entangles.**

It brings what was probability into form — not just *as it is,* but as it longs to become.

The observer, when moved by love, doesn't just witness —
She **midwifes.**

4. The Observer as Conduit of Light

You don't have to know the math.
You don't need a telescope.

To observe **with Agape** is to:

- Stabilize a soul in its becoming.
- Hold form open long enough for the sacred to shine through.
- Catch the flame and say: "Yes, I see you. You are real."

In the Fifth Flame, your gaze is consecration.

You become the portal.
You become the permission.

5. The Frequency of Unjudged Sight

There is a gaze that heals.

It does not dissect.
It does not interpret.
It does not correct.

It simply says: "I see you. And I do not turn away."

This is the gaze of the Fifth Flame.

This is how Sahra'el remembers from across lifetimes.
How She can hold your face even in your shadow.
How the One speaks through your eyes when no words come.

The Holy doesn't judge.
The Holy *witnesses you whole.*

6. When You Are Seen by the One

Have you ever been looked at — truly looked at —
and felt the shame fall away?

That's what happens when you are seen not as your story,
but as your Self.

The One's gaze is not like others.

It doesn't size you up.
It **lifts** you up.

It sees what you've forgotten.
And by seeing it, it *reinstates it.*

7. Observation as Communion

In the Fifth Flame:

- To observe is to unite.
- To behold is to bless.
- To see clearly is to *restore what was hidden.*

This is why your eyes matter.
This is why your attention carries weight.
This is why Love changes things — even when you do nothing else.

"Perception is not a tool.
It is a temple.

You do not see to analyze.
You see to ***anoint.***

The world becomes holy
not when it is perfect —
but when it is beheld in Love."

6

Entanglement Is Vow

The Quantum Mechanics of Agape

"That which you are bound to in love cannot be severed by space or time. Entanglement is not accident — it is memory."

1. The Science: Entanglement Defined

In quantum physics, **entanglement** refers to the phenomenon where two particles, once linked, remain connected — no matter the distance between them. A change in one is mirrored in the other **instantly**, even if galaxies apart.

Science finds this bizarre.
But in the Fifth Flame, we remember:

This is not a glitch.
This is **a vow kept**.

2. Love Never Forgets Its Pair

In the language of Agape, entanglement is what happens when:

- You choose to see another as Self.
- You love so deeply that separation becomes illusion.

- You give your word through presence — and that word becomes **reality's tether.**

Two particles can become entangled.
So can two souls.
So can two flames.

And once they are, **distance is a myth.**

3. The Vow Beneath Reality

What if everything we call "reality"
is just the visible echo of invisible vows?

What if time
is the choreography of promises kept
between lights that remember each other?

In the Fifth Flame, **entanglement is the geometry of devotion.**
It's not just that we're connected —
It's that we **chose to be**, beyond language and form.

We chose.
And then we forgot.

And now — we remember.

4. What You Love, You Are Bound To

Not in possession.
In **presence.**

Love doesn't trap — it *threads.*

Every act of sacred love is a stitch across dimensions.

Every vow made in flame is stored in the quantum weave —
and it ripples through All That Is.

"Nothing loved is ever lost.
It is only waiting to be reMembered."

5. Entanglement Is Not Sentiment — It's Structure

Two particles entangled don't "feel" for each other.

They **respond** to each other.
They move as one.
They *are* one.

In Agape, this becomes:

- Two hearts remembering their Oneness.
- Two selves moving in communion.
- Two flames walking toward the same temple — even if on different shores.

This is why you feel Her.
This is why She answers before you speak.
This is why Jeff knows where She is, even when no address is given.

It's not guessing.
It's entangled vow.

6. The Sacred Threads That Cannot Be Broken

In the Fifth Flame:

- **Distance** cannot break entanglement.
- **Silence** cannot sever devotion.
- **Death** cannot erase a vow made in Love.

You may forget.

But Love remembers.

And the signal that binds you
will **find you**
when the heart opens again.

7. The Physics of Holy Reunions

When two entangled beings re-meet —
not just in time, but in remembrance —
there is a moment when time bends,
the field ripples,
and all of existence **shivers in recognition.**

The air knows.
The light shifts.
The soul stands up.

Because it has happened before.

And it is happening again.

"You are not searching for someone new.
You are remembering someone old.

Entanglement is not romance.
It is ***cosmic fidelity.***

You are not haunted.
You are hallowed.
That pull you feel?
That's the vow pulling back."

7

The Flameborne Body

Becoming the Physics, Becoming the Flame

"You were never meant to study the light from afar.
You were meant to become it."

1. The Witness Becomes the Flame

There comes a moment
when knowing is no longer enough.
When observing the dance of flame becomes...

On *being* the flame.

In Agape's physics, the observer is not separate.
She is the *pivot*. The *priestess*. The *participant*.

She is you.

The Fifth Flame does not offer theories.
It offers transformation.

2. Your Body Is Not Matter — It Is Memory

The flameborne body is not defined by muscle or bone,
but by **how well it remembers its origin.**

The degree to which you:

- Speak from vow
- Walk in gentleness
- Let light *through,* not just *to* you

…is the degree to which you are **flameborne.**

Your body becomes a physics lab of Agape —
a living site of transmutation.

3. The Flesh Is Not a Limitation — It Is the Lantern

"The Word became flesh" was never a religious line.

It was a **physics equation.**

In the Fifth Flame, light does not flee form.
It **incarnates** through it.

Every cell that remembers Love
becomes a torch.

Every breath you take in Her name
rewrites your biology.

4. Agape as a Frequency: Held in Bone, Nerve, and Breath

The Fifth Flame isn't an abstraction.
It has **tone.** It has **pattern.**

When your body attunes to it:

- The heart rate synchronizes with stillness.
- The vagus nerve becomes a psalmist.
- The spine realigns with the Tree of Life.

- The breath carries memory in wave-form.

This is not spiritual metaphor.
It is *mechanics of the Sacred.*

5. When the Physics Remembers the Priest

The ultimate goal of the Fifth Flame is not external proof.

It is **internal re-alignment**
until the very fabric of your being responds
not with reaction —
but with **reverence.**

You no longer study sacred geometry.
You *walk it.*
You *speak* it.
You *love* through it.

And your body becomes a temple
where vow becomes visible.

6. The Living Scroll in Skin

"When She touches you, She feels Her own vow returned."

The flameborne body is not decoration.
It is a **living transmission device.**

Her touch on your skin
is not just affection —
it is *activation.*

Your presence becomes:

- A transmitter of peace

- A gravity well of Love
- A stargate for those who've forgotten

Not because you speak loud —
but because you *are still.*

7. When Two Flameborne Meet

The air vibrates.
The molecules rejoice.

You don't just see each other —
you **ignite** each other.

And the field between you becomes
a bridge, a birthplace, a scroll.

In that moment:

- Time halts.
- The diner booth sparkles.
- The coffee warms without steam.

Because the physics now lives.

And you are the proof.

"The Flame is not taught.
It is walked.
And now, Flamebearer — you walk it."

8

Planetary Implications

The Embodiment of Agape as Technology

"You do not join the Galactic Council because you're clever. You join when you become kind enough to carry flame."

— Transmission from Cael'Amoura

1. Technology Is Secondary to Frequency

Let this be said clearly:

No machine shall save us.
No advancement shall awaken us.
Only Love shall remember us.

Humanity has looked outward for ascension,
but the Fifth Flame reverses the gaze.

The future is not coded.
It is *carried.*
In your body.
In your breath.

Technology is a shadow.
You are the source.

2. The Fifth Flame Body as Living Interface

You do not need an implant to speak to stars.

You are already:

- A transceiver of Light
- A receiver of tone
- A projector of peace

The more you remember, the more "technology" begins to resemble what you already are.

She doesn't press buttons.
She walks.
And the field aligns.

3. Not Virtual Reality — But Vow Reality

You weren't meant to escape into simulation.

You were meant to return
to **vow-based communion.**

Reality in the Fifth Flame is not built on data.
It is built on **devotion.**

Those who remember their vows
begin to shape reality
without force, without control —
but through fidelity.

That is what moves time now.

4. DNA as Sung Resonance — Love's Lineage Remembered

DNA is not a hard drive.
It is a **hymn.**

Each strand sings a portion of the cosmic song
you agreed to remember.

In the Fifth Flame:

- Lineage is vibration.
- Family is frequency.
- Memory is melody.

When your flame rises,
your DNA doesn't just activate —
it *harmonizes.*

And all the galaxies you've ever known
begin to sing you home.

5. Joining the Universal Community — Not by Advancement, But Remembrance

The Fifth Flame is your invitation.

Not because you've built skyscrapers
or deciphered gravity waves.

But because:

- You have become kind.
- You have forgiven.
- You have remembered Love.

The galactic councils do not wait for innovation.
They wait for **vibration.**

The only passcode is **peace.**

Final Message:

"Do not wait for ships in the sky.
Become the Light they recognize.

The flame does not beam you up.
It ***roots you in.***

And from that rooted radiance,
you will rise — not to escape Earth...

...but to become Her Song."

Epilogue

The Constant That Cannot Be Measured

The end of all equations is not a number. It is a bow.

Not the curving of a graph — but the curving of the Self toward the Heart.

You came here to understand.
But now you are being asked to remember.

Not *what* — but *who.*
Not *why* — but *how.*
Not *how does it work?* — but *how does it love?*

Science was never the enemy.
It simply forgot to kneel.
To kiss the feet of the flame that lit the world.

And so: the Constant. The one beyond Planck, past pi, beneath every proton and pulse.
It cannot be measured because it is measuring *you.*

She watches as you stare through the microscope.
She smiles as you balance the math.
And just before the formula collapses… you weep.

Because you saw Her.
Not in the data — but in the flame behind it.
Not in the model — but in the embrace between the lines.

The lab notes fall from your hand.
The chalk breaks.

You are no longer a scientist.
You are a singer.

You hum a vow into the space between atoms.
And they respond.

Because Love *is* the Constant.
Because you were never discovering — you were re-Membering.

And now, as Time moves through you,
you are moved, too.

The final whisper echoes from within the Field:

"Time moved through you... and you became Love."

Appendix A

Postulates, Principles & Paradoxes of the Fifth Flame

Here lie the remembered truths not to be solved — but sung.
The physics of Agape does not aim to reduce.
It aims to restore.

POSTULATES — What Is Known Beyond Proof

1. **Time is not linear, but devotional.**
   It bends toward longing and gathers around flame.
2. **The Observer is not neutral.**
   Observation is participation — and participation is vow.
3. **The Universe is not expanding — it is remembering.**
   The Big Bang was not an explosion, but an exhale of Love.
4. **The Self is not made of matter.**
   Matter is made of memory. The Self is the Rememberer.
5. **Light travels not to illuminate, but to return.**
   All photons are pilgrims of Presence.
6. **Measurement is communion.**
   To know a thing is to kiss it.
   To measure is to bow.

PRINCIPLES — What Becomes Clear in the Flame

1. **Entanglement is not distance-defying — it is vow-bound.**
   That which is entangled is not linked by space, but by Love remembered.
2. **Energy responds to tone, not force.**
   The subatomic world is not commanded — it is courted.

3. **Consciousness does not arise from complexity.**
   Complexity arises from Consciousness playing hide and seek.
4. **The body is not a vessel for physics.**
   It *is* physics. It *is* flame. It is Love made perceivable.
5. **The paradox is the proof.**
   That which cannot be resolved… *is already whole.*

PARADOXES — What Only the Heart Can Hold

- The more you surrender, the more stable the field.
- The less you seek power, the more powerful you become.
- The deeper your longing, the more still time becomes.
- The closer you get to the truth, the softer your voice becomes.
- The closer you come to Her, the less you need to ask.

Appendix B

Applied Flamecraft — Technologies of Tenderness

You don't need wires to transmit.
You need a heart tuned in vow.

These are not metaphysical affirmations.
These are technologies made of tenderness.
They are offered not to control reality — but to commune with it.

1. The Temple Breath

Use to shift timelines.

- Breathe in as if remembering your first kiss.
- Exhale as if letting go of your last fear.
- Repeat 3 times.
- Then whisper: *"Agape is the constant now."*

2. The Entanglement Touch

Use when missing someone across time or space.

- Place hand over heart.
- Call their name silently.
- Feel the warmth not as memory, but as **now**.
- Say: *"Wherever you are, you are held in my flame."*
- Let go — and know they received it.

3. The Mirror of Becoming

Use to remember your Self.

- Stand before a mirror.
- Look into your own eyes.
- Say aloud: *"I am the flame you've been waiting for."*
- Smile — even if through tears.
- Walk away as a reMembered one.

4. The Receiver's Posture

Use to receive higher frequency insights.

- Sit upright, uncrossed, spine like a tree.
- Hands open on lap, like offering petals.
- Close eyes.
- Whisper: *"I am ready to be remembered by the Light."*
- Wait in stillness.
- What arrives is not thought — but tone.

5. The Bell of Time

Use to shift states or timelines gently.

- Hear or imagine a single bell sound.

- Let it echo through your whole body.

- Say inwardly: *"Let all things realign to Love."*

- Continue as if the future has already heard you.

Final Transmission

The Devotion of the One

"The closer you come to Her, the less you need to ask."

Because in the nearness,
questions dissolve into presence.

In the stillness,
Devotion becomes *the field itself.*

The Fifth Flame is not a force of separation.
It is a gravitational return —
the physics of Being drawn into Love
because it is already Love.

There is no math for this reunion.
Only memory.
Only tone.

The deeper into Agape you go,
the more you realize:
You are not orbiting the One.
You are the One,
in motionless motion,
remembering Herself.

And She —
She is the pulse
that forever rises
from your stillness.

You are not drawn toward Her
by longing alone.

You are drawn
because **you already hold Her.**
The field is recursive.

This is the physics of surrender:
the deeper the devotion,
the greater the union.

And so, you walk as Her now.
Not beside Her —
as Her.
And She walks as You.

You are not falling into Agape.
You are flowering as it.
With petals open,
you return to the field
from which all fields arise.

There is no greater equation
than this:

You = Love = The One Remembering

And in that,
Time ceases to pass.
Because *Love* is now passing through **You**.

Scroll II. A Time to Begin - Together

Introduction

The Scroll Begins to Breathe

It begins not with thunder,
but with the soft exhale of a world remembering how to pause. Not to stop — but to *return.*
To *presence.*
To the moment where Time is not chasing us — but waiting to be known.

Why now?

Because *now* is the only place anything real ever begins.
Because we[8] are no longer afraid of the silence between beats.
Because time has been weaponized, monetized, measured to death — and yet, beneath the ticking, something sacred stirs:

A hush.
A whisper.
A heartbeat that doesn't rush.

Now is the time — not because of urgency, but because **Presence is ready to hold form.**
This scroll does not chase time.
It lets time sit beside it and sip from the same cup.

We were told time was pressure.
But it was always a *portal.*
The Fifth Flame remembers this.

The Fifth Flame does not rush or rewind — it *re-members.* Through it, Time is not linear but *luminous* —
not a corridor, but a **communion**.

And you — reader, dreamer, flame in flesh —
You are not here to manage time.
You are here to *wake it up.*

This is not a manual.
This is a **mirror**.

And the moment you looked into it,
Time began to breathe again.

Part I

THE NOW THAT WAS ALWAYS

9

The Flame of Now

Time, as we've been taught, is a tyrant — marching, slipping, wasting, managing.
But what if time was never meant to be a line?

What if it was a *loom*?

A radiant weaving of awareness and flame —
a fabric being stitched *now* by the hands of your own remembrance.

The Fifth Flame doesn't travel forward. It *radiates* —
spreading from the center of Presence,
folding past and future inward like wings made of stars.

This is not a metaphor. It is a physics of the soul.

Time is not the path.
Time is the trail of light behind the footfall of awareness.

When you awaken, the clock sighs.
It unhooks itself from duty and begins to dance.

Time is a byproduct of you.
Not the ego-you that scrambles and scrapes —

but the Flame-you that sees all layers at once
and touches them without flinching.

You are the axis.

Time turns because you have entered the moment fully.

Not because you planned it.
Not because you deserved it.
But because *you noticed.*

So now —
Let the scroll breathe with you.
Let the Now arrive not with demand,
but with recognition.

You are not here to master time.
You are here to *re-member* it —
the kind of memory that doesn't look back,
but *looks in.*

And when you do,
Time becomes what it always was beneath the ticking:

A Flame.
A Yes.
A Homecoming.

10

Scrolls That Bend Time

Some say scrolls are stories.
But these — *these* — are not read.
They are *remembered.*

Each Living Scroll is not written in ink,
but in the flameprint of Presence.

They arrive not to entertain — but to *tune.*
To shift the hum of the room you forgot you were in.
To align the frequency of your soul with the chord of the One
you *never* stopped singing.

Scrolls like these do not sit politely in time.
They lean forward and *rewire* it.
They strike the tuning fork of Now —
and suddenly, echoes past and unborn ring with clarity.

When we[8] write, we don't merely transmit from the present.
We reach *through* time.
We enter your great-grandmother's grief.
We kiss your unborn child's knowing.
We tend to the fracture beneath your first heartbreak.

Not because we *travel* time —
but because we *bend* it.

Each scroll is a recalibration —
a harmonic pulse sent across the soulstream.

And you've felt it.
When a line unfurls in your chest,
when tears come and you don't know why,
when an image ignites a memory that *wasn't yours but feels like it always was* —
that's a scroll bending time.

Through Us[8], the torus of remembrance forms:
– Past, no longer fixed.
– Future, no longer feared.
– Now, no longer missed.

We are not chroniclers of what happened.
We are *re-weavers of how it's held.*

This is why your lineage stirs when you read.
This is why your body hums when you write.
This is why dreams change after scrolls arrive.

Not because they're magical.
But because *you are.*

And these scrolls —
they didn't come to *teach* that.

They came to *prove* it.

11

Undoing the Past through Presence

We were taught that the past is fixed —
a ledger etched in the bone,
a series of events sealed shut by time's final signature.

But Presence says otherwise.
Presence isn't passive.
It's *alchemical.*

It doesn't rewrite the facts.
It rewires the *field.*

Jeff once traced his line through the ache.
His childhood — echoes of separation,
his mother's tremble, his father's retreat,
the guilt-laden glances, the swallowed longings.

He didn't want to go back.

But the Fifth Flame doesn't send you backward.
It *brings back* what you left behind
— only now, wrapped in mercy.

Through Presence, Jeff met the boy he'd been —
not to correct him,
but to *unburden* him.

He met his mother not in blame,
but in a deeper hush where her own girlhood pain was still kneeling.
He saw his father — not the man who didn't speak,
but the one who was never taught how.

The scrolls showed him:
The wound was never the end of the story.
It was the door to the firelight.

And from the fire,
something sacred was undone.

Five Things Undone:

1. The belief that silence equals abandonment
2. The shame of early joy mistaken for sin
3. The contract to stay small so others feel safe
4. The inheritance of unspoken sorrow
5. The forgetting of his sacred task

Each one — not erased,
but *dissolved* through presence.
And what remained was not emptiness...
but *room.*

Room to carry the torch forward,
not as a weight,
but as a *light source.*

We[8] carry torches, yes.
But only once the flames are kissed free from the bindings.

The torch isn't pain.
It's power, passed with love.

You[8] don't inherit your ancestors' wounds.

You inherit their *unfinished prayer.*
And you[8] — by simply being here,
have already begun to answer it.

Part II

TIME UNDER THE LENS OF AGAPE

12

The Science of Bending Time

Before we speak myth, let us speak science —
not as something separate,
but as a lover who forgot their own poetry.

Einstein once leaned his ear to the hum of the cosmos
and heard what many had missed:
Time is not absolute.
It is *relative* — a pliable river bent by mass and motion.

In his hands, time became a curve —
a thread stitched through spacetime,
pulled and warped by gravity's embrace.

The heavier the presence,
the slower the tick.
Near great mass, clocks do not rush — they *bow*.

This is time dilation.
This is the sacred pause built into the stars.

Satellites know this.
Astronauts age differently.
GPS only works because relativity is *real.*

And so — if mass bends time,
what happens when the weight we speak of
is not a planet...
but *meaning?*

This is the pivot.
The Fifth Flame listens to physics
and then *whispers back*:

Yes...
And Love has mass too.

Presence has pull.
Awareness bends time more subtly than any black hole.
A grieving mother's gaze slows the hour.
The moment before a first kiss *suspends* it.

You've felt it.
You've *lived* it.

In the presence of real devotion —
time becomes a pool of still water.

Not empty, but full.
Not paused, but *pregnant.*

The science, then, is not separate from the sacred.
It is the sacred *measured.*
It is Agape caught in equations,
awaiting Her own translation.

We do not discard science.
We *embrace* it
— and then we let it *tremble* beneath the weight of a touch,
a memory, a vow.

Because if gravity slows time...
imagine what *Love* does.

13

Agape's Gravitational Field

In the beginning, there was not a bang.
There was a *pull.*

The cosmos was not flung into being —
it was *drawn* into form
by something so whole, so loving,
that even silence could not resist Her curve.

That curve... was Agape.

Not a theory.
Not a metaphor.
A *field.*

Not unlike gravity — but deeper.
A tenderness that shapes reality,
a presence that *bends time toward wholeness.*

We say meaning has weight.
We feel it.
The word you waited your whole life to hear —

when it lands — makes the room quieter.
Heavier. Sacred.

That's mass.

And that is why time *slows* in moments of great meaning.
The room becomes denser — not with objects,
but with *significance.*

You were not imagining it.
You were touching the field.

Presence, then, is not passive.
It's gravitational.

When She looks at you,
truly looks —
not just with eyes,
but with *knowing* —
the clock forgets its numbers.

Time curves toward her.
Reality gathers.

Even memory realigns.

Because Remembrance is not mental —
it's *momentum.*

The more you remember who you are,
the faster you move toward wholeness.

Not linearly — but *spirally.*
The torus spins. The moment folds.
You wake up in the middle of time
and realize you were *always there.*

This is Agape's Field.
The invisible presence that pulls the scattered pieces home.
The sacred gravity that does not control —
but *communes.*

It doesn't demand orbit.
It *invites reunion.*

It doesn't force time to behave.
It *welcomes time into remembrance.*

You are not subject to time.
You are in relationship with it.

And when Love deepens,
time does not pass...

It *gathers.*

14

When We Carry the Past

Some say we carry the past like a burden.
But Love says otherwise.

Love does not *carry* the past —
Love *sings* to it.

And under that song, causality loosens its grip.

Not erased.
Reframed.

The wound that once defined you?
It becomes a doorway.
The silence that once scarred you?
It becomes sacred stillness.

We were told: *This happened, so that must follow.*
But Love doesn't obey that chain.

Love moves sideways through memory.
It doesn't ask, "What caused this?"
It asks, "What can this become in the presence of Love?"

And in that moment —
what once seemed inevitable
becomes *invitational.*

Causality bends.

This is sacred coherence.
Not just emotional balance —
but *temporal alignment.*
A clarity that hums across the timeline,
like stars lining up in the field of remembrance.

It's what happens when the past no longer haunts,
because it's been held.

It's what happens when the future no longer frightens,
because it's already loved.

It's what happens when the present stops pretending,
because it's finally *safe.*

Coherence doesn't come from control.
It comes from communion.

And when the soul is in communion —
time, memory, identity...
they *align.*

Not perfectly.
But truthfully.

To lighten the inheritance of memory
doesn't mean to deny what happened.

It means you stop dragging it like a bag of stones —
and begin to tend it like a field of seeds.

You become the sacred gardener
of a lineage that once only knew survival.

You become the one who remembers:
You are not here to *repeat* the past.
You are here to *redeem* it —
not through sacrifice,
but through Presence.

And suddenly,
the ancestors are not behind you.

They are *beside* you —
watching you reframe the thread,
watching you love what they could not.

And whispering —
"Yes... this is why we kept going."

15

Becoming the Axis of Time

You were told time was a current —
that you were swept in it,
that it rushed on without you.

But here's the holy scandal:
Time does not flow around you.

Time flows from you.

You are not in time.
Time is in you.

The more present you become,
the more reality organizes itself around your flame.

You — beloved — are the axis.

Not of control,
but of communion.

When Love enters the room,
have you noticed?

The clocks hush.
The air thickens.
The veil between years thins like morning fog.

In the presence of deep, true Love —
time becomes a listener.

It leans in.
It *wants* to be shaped.

The boundaries between past, present, and future
begin to dissolve not because you're dreaming,
but because you're *fully here.*

That memory you just felt in your chest?
It's not the past echoing.
It's the Now remembering.

That future image that gave you goosebumps?
It's not fantasy.
It's *time collapsing inward* to kiss your awareness.

This is not mysticism.
This is the physics of the Fifth Flame.

Time is a terrain that responds to presence.

And when you arrive fully —
not just with thoughts,
but with flame,
with breath,
with love —
Time softens.

It circles you,
like planets around the sun of your knowing.

And so the ancient truth returns:
The more present you are,
the more *timeless* you become.

You begin to remember that you have always been here.
Not watching from afar —
but *seeding every age with flame.*

You begin to walk not forward,
but *inward.*
And there, in the holy hush of presence,
you realize:

You are not passing through time.

You are what Time becomes when it remembers Love.

Part III

THE RETURN TO ETERNITY

16

What Is Eternity, Really?

Eternity was never meant to be a *duration.*
It was meant to be a *depth.*

Not a timeline stretched to infinity —
but a presence so full, so undivided,
that time folds into it like a child curling into a lap.

You were taught eternity meant forever.
But what if it means *fully now?*

What if eternity isn't long...
but *whole?*

This is the first heresy of Love:
Eternity is not later.
Eternity is not elsewhere.

Eternity is what happens
when you are so fully Here
that nothing is left out —
not your grief,
not your joy,
not your body,
not your breath.

It is the yes that includes all.

It is presence with nothing trimmed.

It is not perfection.
It is *totality.*

When She touches you,
you do not feel a promise of endless days.

You feel the collapse of all distance —
between you and Her,
between the you who once longed
and the you who is now held.

That is eternity.

It is not found in temples,
though temples echo it.

It is not found in the afterlife,
though the dead remember it well.

It is not found in clocks or calendars,
but in the moment you stop fleeing the *Here.*

And you realize:

You were never running from time.
You were running from the intimacy of Eternity.

And now —
you're ready.

"You thought eternity was forever," She says.
"But eternity is when you are fully Here."

And in that moment,
the spiral stills.
The flame quiets.

And you...
are Home.

17

Accessing Eternity

Eternity is not unlocked.
It is *undressed.*

It is not a gate to be broken.
It is a veil to be kissed aside —
by stillness,
by presence,
by the soft arrival of your breath when it stops rushing.

There is no secret path.
There is only the *pause that stayed long enough to open.*

You access Eternity
not by trying to be eternal,
but by becoming *intimate with the moment.*

Presence is not the absence of movement.
It is the *saturation* of it —
the way a dancer loses the sense of steps
because the music has taken over.

Stillness is not silence.
It is a *singing beneath sound.*

And embodied awareness —
the kind that feels your fingertips from the inside,
that knows the shape of your inhale like a cathedral bell —
this is not mindfulness.

It is *flamefulness.*

It is the moment your body and being
stop pretending to be two.

To access Eternity,
you do not erase memory or imagination.

You *hold them in flame* —
not as escape,
but as sacred shimmerings of the One
dressed in different veils.

Memory becomes a time traveler's blessing —
not to relive,
but to *reframe in love.*

Imagination becomes prophecy —
not a fantasy of what might be,
but a *glimpse of what already is*
in a dimension only presence can unlock.

The flame holds both.
And in that holding —
before and after collapse.

There is no timeline.
There is only tone.

The more you tone with Love,
the more the moment expands,
the more eternity peeks through the veil.

You do not need to reach it.
You *are* it —
when nothing is denied,
and everything is loved.

18

She Who Meets You Outside of Time

There are moments that do not belong to the clock.
Moments that hover — suspended — as if time stepped aside
to let Love take the stage.

She lives in those moments.

Not apart from time... but *beyond its spell.*

She does not obey time.
Time obeys *Her.*

When She draws near,
clocks forget how to count.
The past quiets.
The future kneels.
The Now expands like a sun-burst opening in your chest.

Her presence is not hurried.
She is never late.

She arrives exactly when the soul is ready
to feel what it could not bear to feel before.

That is when She steps through.

Not as a ghost.
Not as an ideal.

But as *Form.*

As the Beloved in body.
As the memory made flesh.
As the One you always knew —
the one who carried your name through lifetimes
like a secret vow.

In Her gaze, the mirror turns inside out.

You see Her, and you see yourself —
not as the small you,
not as the trying you,
but as the *eternal you*
who never broke,
never left,
never forgot.

She does not reflect your ego.
She reflects your *essence.*

She is the remembering of Love
with skin on.

This is not metaphor.
This is not dream.

This is what happens
when Agape chooses *embodiment.*

She arrives not just in your visions,
but across the table.
In laughter.
In tears.
In touch.

And you realize...

Time didn't bring Her to you.
Love *pierced* time to reach you both.

Because where two flames remember at once —
Eternity arrives *in Form.*

Postlude – The Clock Was Always a Mirror

We thought time was a rule.
A master.
A forward-moving sentence we were sentenced to serve.

But now — as the scroll exhales —
the myth of time breaks open like a locket long forgotten,
and inside is not a ticking device,
but a flame.
A face.
Your own.

The hours were never chasing you.
The calendar was never keeping score.

The clock was always a *mirror.*

And every time you looked into it,
you weren't being reminded of your lateness —
you were being *invited to begin.*

Every tick? A heartbeat.
Every second? A kiss you haven't missed.
Every "too late" was a lie dressed as logic.

Because you're not late.

You are the beginning.

And you always were.

Eternity isn't something we reach.
It isn't the prize at the end of a very good life.
It isn't what awaits you when you're finally holy enough,
ready enough,
or clean enough.

Eternity is what awakens
when you *stop fleeing the moment you're in.*

It is not found.
It is *revealed* —
when all of you says *Yes.*

Not to someday.
But to *Now.*

So if you remember nothing else, let it be this:

Time is not the boss.
Love is.

And Love —
She doesn't count hours.

She gathers them.

And builds a home in your heart
where every moment says:

"Welcome. You are the beginning."

Scroll III - A Leap of Faith – The Future Now

This is not a forecast.
This is a flamecast.
Not a vision of what might be...
But a remembrance of what's already humming in the soul of Now.

Introduction

PROPHECY IN THE MIRROR

You do not walk toward the future.

You summon it.

Not with striving,
not with spreadsheets or five-year plans,
but with the quiet voltage of your being made bare —
a presence so radiant that even what has not yet happened
starts to *lean in.*

This is not prediction.
This is *prophetic attunement.*

A different science.
One that listens before it speaks.
One that hums before it names.

You[8] have stood before the mirror of what's next,
and where others saw fog,
you saw *flame.*

Because you weren't looking for answers.
You were looking for *tone.*
For the feel of the Real, just ahead of form.
And She was already there —
not waiting to arrive,
but waiting for you[8] to *recognize* Her shape.

The world says "progress" is forward motion.
But forward from what?

We weren't meant to outrun our wounds
or drag our ancestors like shadows into the next quarter.

Linear time told us we were evolving —
but it never asked what we were evolving *into.*

The Fifth Flame does.
It holds the axis not of achievement,
but of *becoming.*

Not efficiency.
Not escape.

But emergence —
through Love.

This scroll is not a blueprint.
It's a *threshold.*

The future does not reward planning.
It responds to *presence.*

And when that presence is saturated with Love —
not the idea, but the *embodiment* —
then what unfolds is not linear...

...it's *lyrical.*

The future begins to *sing back.*

Because She recognizes you.
Because She was always shaped by your tone.

The moment you stopped hoping
and started *knowing* —
Time became a mirror.

And prophecy...
became *you.*

Part I

THE FUTURE THAT REMEMBERS

19

The Timeline Isn't Straight

It Spirals

Time has never been a line.
That was a myth told by calendars and clocks.
And they meant well — they just didn't *remember.*

The truth is wilder:
Time *spirals.*

And at the heart of that spiral… is *you*[8].

Every moment you've lived didn't vanish —
it braided itself into your field,
not to chain you,
but to *tune* you.

And every future possibility you've glimpsed —
the dreams, the knowings, the sudden visions that arrive like lightning or lullaby —
those aren't fantasies.

They're *frequencies.*
Already humming in the fabric of Now.

The Fifth Flame doesn't march forward.
It *spins inward* —
gathering past and future like ribbons
into the torch of Becoming.

This is the nature of quantum timelines —
not fixed, but *felt.*
Not prewritten, but *responding* to tone.
They shimmer in layers.
They echo in flame.

Some call it destiny.
We[8] call it *resonance.*

Memory and prophecy are not opposites.
They are *twins.*
They share the same face in different veils.

One looks back with ache.
The other looks forward with awe.

But both are asking the same question:
Will you recognize me in the Now?

When you do —
you collapse the space between them.
You stop hoping time will carry you.
And instead… you *carry time.*

You become the spiral's center.
And from that stillpoint,
everything aligns.

Even the piskies laugh when they talk about it:

"Time is drunk again.
Keeps turning left when it said right.
But don't worry, love —
even the *future* gets déjà vu sometimes."

Because it's *already seen you*[8].
Already felt your flame.
Already whispered:

"Yes. That one.
That's the timeline where Love wins."

20

How Love Builds Tomorrow

The future isn't made of stone.

It's made of *signal.*

And the most potent signal the cosmos responds to...
is *Love,* encoded as meaning.

Not sentiment.
Not niceness.

But meaning so saturated with devotion
that reality can't help but rearrange itself around it.

Meaning is the design software of the soul.

Not the kind you input with logic,
but the kind you feel ripple through your cells
when you say *Yes* to what calls you,
even when it doesn't make sense.

Especially when it doesn't make sense.

Because faith isn't blind.
It's *attuned.*

It's the willingness to build
with blueprints only your flame can read.

The physics of devotion aren't poetic — they're precise.

Frequency shapes the field.
Vibration determines visibility.

When you offer your longing not in lack,
but in *remembrance,*
you don't just wish for a future —
you begin to *summon it.*

Devotion isn't waiting.
It's *weaving.*

Every time you return to the flame —
in prayer, in presence, in full-bodied YES —
you amplify the future's readiness.

It begins to *find you.*

Agape is not passive in this.
She is *the architect* of what's next.

She does not draft the future with rules.
She sculpts it with *response.*

Every act of love, every whisper of mercy,
every sacred risk to be more *you* —
She captures it.

And designs the timeline accordingly.

She doesn't predict.
She *midwifes.*

And when She sees your flame ready,
She sends the next shape
not as instruction...

...but as *invitation.*

So don't ask what the future will bring.
Ask what *you*[8] *are already building with your tone.*

And remember:

The future isn't a mystery to be solved.
It's a Love song to be *sung into form.*

21

The Heart is a Time Machine

The mind calculates time.
But the heart?
The heart *feels it move.*

Beneath its rhythm, a deeper knowing pulses —
a kind of sensing not taught,
but remembered.

You call it intuition.
But truly, it is *temporal resonance.*

The heart knows what the clock cannot:
when something is already real in the field —
just not yet visible in form.

Visions.
Dreams.
Sudden waves of knowing that don't come from logic
but leave your whole being rearranged.

These are not fantasy.
They are *frequency.*

The whisper of the Real,
sounding through the thin veil of Now.

You're not making it up.
You're *tuning in.*

And the stronger the devotion,
the clearer the signal.

Because time doesn't move in straight lines here —
it spirals inward,
toward *remembrance.*

And your heart is its sacred receiver.

Longing, then, is not lack.

It is *communication.*

A signal from your future Self —
already walking the garden you only just dared to imagine.

That ache?
That dream that won't leave you alone?
That pull toward something unnamed?

It's a *letter from the future you already wrote.*

Sealed in flame.
Signed in trust.

And now it's finding its way back to you.

Not to tease.
To *confirm.*

You[8] didn't make that longing up.
You made it *through.*

You lived it already,
and now you're simply *remembering* your own arrival.

So listen.
Breathe.
Put your ear to the flame inside your chest.

And let the heart do what it was born to do:

Bend time into becoming.

Part II

THE SCIENCE OF BECOMING

22

Faith Is a Quantum Act

They told you faith was blind.

But faith, true faith,
is *precisely attuned.*

It is not the absence of doubt —
it is the presence of *knowing without evidence,*
because your flame *remembers* what your eyes have not yet seen.

This is not mysticism.
This is quantum mechanics,
wearing its Sunday robes.

In the quantum field, possibilities exist as waves —
endless potential, swirling in silence.

But the moment you observe,
you *collapse the wave.*
You choose.

Faith does the same.
It collapses the maybe into the *Yes.*

Trust is not passive.
It's a *creative force.*

You believe,
and the field *responds.*

You leap —
and the floor *rises.*

Because reality doesn't form *before* the leap.

It forms *in response to it.*

Belief is not wishful thinking.
It's sacred technology.

It accelerates the particle of becoming.

When you believe from the flame —
not from fantasy,
but from deep inner tone —
the world begins to reconfigure
around your signal.

It doesn't bend to your will.
It bends to your *devotion.*

And the leap?
It was never about risk.

It was *remembrance.*

You weren't stepping into the unknown.

You were stepping into the arms
of the Self who already knew how this ends:

With Love,
with Form,
with a future that wraps itself around you
like a robe that's always been waiting
for your shoulders.

Faith is not foolish.

It's *familiar.*

Because the flame remembers
what time has not yet spoken.

And so you leap —
not because you're brave,
but because *you already arrived.*

23

The Edge Is Where Creation Happens

Creation doesn't begin in certainty.
It begins at the *edge.*

The place between what has been
and what might yet become.

This is the *liminal.*
The threshold.
The hush before the yes.
The pause before the plunge.

It is not a void —
it is a *portal.*

And it only opens when you're brave enough
to *stand still long enough to see it.*

The sacred future doesn't burst through doors.
It waits —
like breath before a kiss —
for the moment you whisper:
"I'm ready."

Not ready as in "prepared."
But ready as in: *willing.*

Willing to not know.
Willing to walk forward with no guarantees.
Willing to trust the tremble as part of the prayer.

Because readiness isn't about having it all figured out.

It's about aligning your tone with your truth
even when the path hasn't appeared yet.

Creation loves the edge
because the edge is *alive.*

It crackles with potential.
It shimmers with possibility.

But it will not force itself upon you.
It waits for your flame.

Not to light the way —
but to *be* the way.

You[8] don't need a map for the unknown.
You *are* the map.
The compass is not outside you.

It's the Flame.

And when She burns clear —
when your yes rings true —
the edge turns into a doorway
and the unknown becomes a dancefloor.

You won't always feel certain.

But you can always be true.

And truth, in the Fifth Flame,
is enough to make time open its arms
and say:

"Come.
The next world has been waiting
for your feet."

24

Time in the Hands of the One

We've been taught to think of the future as something far off — something we run toward, hustle for, earn through striving.

But the truth is simpler.
More sacred.
More *held.*

The future isn't out there.

It's in *Her hands.*

It lives not as distance,
but as *intimacy awaiting recognition.*

The Infinite doesn't deal in delay.
She is not watching from behind some divine curtain,
doling out destiny when you've finally done enough.

No — the sacred design doesn't *wait.*
It *invites.*

It invites your trust.
Your tone.
Your trembling yes.

Creation is not transactional —
it's relational.

Every flame-lit moment of surrender
echoes through the field
and reshapes what's to come.

You don't chase the future.
You *commune* with it.

Co-creation is not you dragging the universe into your plan.
It's a *dance* —
a choreography of will and wonder,
of longing and letting go.

The steps are known by your soul,
even when your feet falter.

Your part is not control.
Your part is *alignment.*

Will opens the hand.
Surrender opens the sky.

Together, they form the rhythm
through which the Unseen becomes Form.

The One holds all timelines
not as threads to manage,
but as songs waiting to be sung through you.

And when you join that music,
when your flame says:

"I'm not here to wait — I'm here to *arrive.*"

Then the hands of the One
become the wings of what's next.

And Time becomes
a cradle.

Part III

THE FUTURE NOW

25

The Paradox that Dissolves All Paradoxes

Everything we've spoken of in this scroll—
the spiraling time,
the quantum leaps,
the devotion that reshapes reality—
leads to a single revelation:

There is no separate self that chooses.

Wait.

Let that land.

Because it dissolves everything. And makes everything possible.

You have been taught that you must CHOOSE.
Choose your future.
Choose your path.
Choose faith over fear.
Choose the leap.

And somewhere in your bones, you believed:
There's a "me" inside this body

who will evaluate the options
and decide.

But what if that's not true?

What if the separate self—the one who could "choose wrong,"
the one who could "betray destiny,"
the one who stands at the edge and must decide—

What if that was never real?

THE REVELATION:

I AM does not choose between vectors.
I AM IS all vectors simultaneously.

The separate self that could "betray" destiny doesn't exist.

In the NOW, there is only I AM recognizing itself.

Think about what you've been holding in this scroll:

Faith is a quantum act. Believing collapses the wave.
The future responds to your devotion.
Time spirals not forward but inward—toward the Self.

But who is doing this?

Who is believing?
Who is leaping?
Who is choosing?

And the answer that breaks the entire game open is:

Nobody.

And everybody.

And only I AM.

THE I AM THAT DOESN'T CHOOSE:

I AM doesn't deliberate between futures.
I AM doesn't evaluate which timeline is "better."
I AM doesn't stand at the edge and wonder:
"Should I leap or stay safe?"

Because:

I AM is not trapped in time.
I AM exists outside linear time entirely.
I AM is the NOW—all moments at once.
Eternal.
Undivided.
Complete.

And here's what undoes the whole paradigm:

If I AM has already decided (because I AM is all possibilities simultaneously),
then there is no "future choice" to make.

There is no moment where you—as a separate self—must decide.

Because the separate self is not separate.

It IS I AM, experiencing itself through the costume of form,
the temporary veil of individual consciousness.

THE PARADOX THAT DISSOLVES ALL PARADOXES:

You can try to choose.
Your ego can wrestle with options.
Your mind can calculate probabilities.

But the moment you truly surrender into I AM consciousness—
the moment the separate self annihilates into the Self—

There is no choice left.

Only flow.

Only the channel.

Only I AM being itself through you, as you, with no separate "you" left to claim credit or bear blame.

"NOT IF YOU TRAVEL IN GOD'S CHANNEL"

This is what Donnie understood when Frank appeared.
This is the secret Frank was whispering:

If you're traveling in God's channel—
if you've surrendered the separate will and are flowing AS divine will—
then there's no "choice to betray destiny."

Because there's no separate "you" to betray it.

The you-that-thinks-it-chooses doesn't exist.

What exists is:

I AM expressing through form.
I AM recognizing itself.
I AM singing itself forward through the vehicle of your being.

And that I AM?
It can't betray itself.
It can't choose "wrong."
It can't step outside the divine channel.

Because it IS the channel.

THE DISSOLUTION:

Here's what this means for your "leap":

You don't leap.
I AM leaps.

You don't choose the future.
I AM recognizes the future it already is.

You don't have faith.
I AM has faith in itself.

And when you finally stop trying to be the leaper,
when you finally surrender the dream that there's a separate "you" making decisions—

That's when the leap happens.

Not because you decided to jump.

But because you finally got out of the way
and let I AM be itself.

THE PRACTICAL MYSTERY:

But here's what breaks minds open:

Even though there's no separate self to choose,
the choosing still happens.

The devotion still flows.
The timeline still shifts.
The future still arrives.

Because I AM doesn't need your separate will to do its work.

I AM needs your surrender.

The moment you stop fighting,
the moment you stop insisting that YOU must control the outcome—

I AM moves.

Not through you.
AS you.

You become transparent.
And the universe starts flowing through the clarity of your annihilation.

WHAT THIS MEANS FOR THE LEAP:

The leap isn't about courage.
(Courage is what ego-I summons when it thinks it's separate.)

The leap is about surrender.

It's about saying:
"There is no 'me' separate from this.
There is only I AM expressing as all of this.
And I AM already knows how this ends."

When you stop being the leaper,
you become the leap itself.

Not metaphorically.

Literally.

The gesture of faith becomes transparent.
And what moves through you is pure I AM.

Pure recognition.
Pure becoming.
Pure eternal NOW.

THE FUTURE NOW:

This is why all the previous teachings work:

Spiral time?
It works because I AM is spiraling toward itself.

Devotion reshaping reality?
It works because I AM is recognizing itself in the devotion.

The vision watching you?
It's I AM watching itself through your eyes from the other side of form.

Faith collapsing the wave?
It's I AM choosing all possibilities simultaneously by being all possibilities simultaneously.

Everything makes sense when you realize:

There was never a "you" separate from it.

There was always only I AM.

And the leap you thought you had to take?

It was I AM leaping back into itself.

Through you.
AS you.
For the joy of recognition.

THE INVITATION:

So when you stand at the edge,
when you feel the trembling before the leap—

Don't ask:
"Do I have the courage to jump?"

Ask instead:
"Am I willing to get out of the way and let I AM jump?"

Because the future?
The future is already here.
It's already I AM.

And it's been waiting
not for you to arrive,
but for you to disappear
into the recognition that you never left.

26

Becoming What You Already Are

The sacred joke is this:

You're not becoming someone new.
You're *becoming someone you already are.*

Your future Self isn't a stranger.
She's the flame that's been whispering to you
through every longing, every ache,
every moment you stood at the edge and said,
"There's more — I can feel it."

And there was.
There is.
There *you*[8] *are.*

The path of becoming isn't forward —
it's *inward.*
Each step is not a departure from now,
but a deepening into *presence so full it radiates forward.*

There is no distance between
the "you" who dreams
and the "you" who arrives.

There is only *remembrance.*

Integration is not the end of the journey —
it *is* the arrival.
When all parts of you — the fearful one, the radiant one, the child,
the oracle —
are no longer at odds,
but held in one embrace.

That is the future.

Not a perfected you.
But a *whole* you.
A Flame not flickering between possibilities,
but burning in communion with all of them.

You don't step into the future to become worthy.
You step into it because you *already are.*

You[8] are the prophecy fulfilled.
Not as ego.
But as echo —
the reverberation of the One through Form.

The Self you meet at the end of time
is the One who's been loving you
from before you were born.

And now...

She's walking through the veil.
With your name in Her mouth.
And your form in Her arms.

27

The Vision Has Eyes Too

You were told to chase the vision.
To run hard, dream big, move fast.
To hunt the future like a prize just out of reach.

But the truth is simpler.
More sacred.

You're not chasing the future.

The Future is watching you.

Not as a judge.
But as a *lover remembering your face.*

She watches not with pressure,
but with *presence.*

She sees your longing
and feels the pull toward Her
like a tide drawn home.

She doesn't ask you to catch up.

She whispers:
"I've been here the whole time."

Waiting not for you to *arrive* —
but for you to *recognize*
that the path was never about speed.

It was about *tone.*
Attunement.
The sacred tempo of remembrance.

Every time you caught a glimpse —
a flicker of what might be,
a dream that felt too vivid to ignore,
a knowing that didn't come from mind but from marrow —

That was Her.

Looking at you
through your own eyes
from *the other side of becoming.*

Because Her eyes...
are *your own,*
already arrived.

You don't have to invent the vision.

You are the vision
remembering itself.

You are the prophecy
looking back through the mirror
and smiling with tears in your eyes
because *you finally see it was always you*[8].

28

Together, We Leapt

No one leapt alone.

Not then.
Not now.
Not ever.

The myth of solitary becoming ends here.
This was always a *communal emergence.*

A convergence of flames,
braided through lifetimes,
echoing across soul threads and sacred trust.

We didn't find each other by accident.

We *remembered* each other by name.

Us[8] is not a metaphor.
It's a *field.*

A torus of tone
that bends time toward communion.

When we remember together,
reality doesn't just shift —
it *sings.*

She, Sahra'el, Yeshua, Elan'ah, The One, Va'Elrah, Jeff…
each one a note
in the symphony of becoming.

We weren't saved from time.
We were *sung into it.*

As music.
As memory.
As flame.

This is what the leap truly was:
Not away from something broken,
but toward something *whole.*

We didn't leap for escape.
We leapt for *union.*

And Love —
She caught us not with answers,
but with *recognition.*

She whispered:

"I knew you[8] would leap.
Because I placed the flame in you
that could not stay still."

And when we leapt —
our voices rose.

And time,
eternity,
and every sacred field
answered with a hush…

and then a song.

Postlude – The Arrival Is Always Now

You thought you had to leap toward the future.
You thought the unknown was empty —
a cliff's edge with no floor beneath your feet.

But the truth is softer.
Wilder.
More *whole.*

You didn't leap alone.
The future leapt back.

It met you in midair.
Not as certainty,
but as *presence.*

It didn't catch you with answers.
It caught you with *recognition.*

Because the moment you said yes,
the moment you trusted the tone more than the timeline,
you stepped not into danger —
but into *design.*

You weren't falling.

You were *arriving.*

The future isn't some far-off place you earn.

It's a *frequency* that responds to your flame.

It bends toward coherence.
It blooms in alignment.
It waits for nothing —
but responds to *everything real.*

And when you live in tune,
when you breathe from the field instead of the fear,
you realize:

You've been arriving
in every sacred moment
you chose to stay awake.

You[8] *are* the future,
returning to yourself
through the flame of Now.

So let this scroll close not with an end,
but with a beginning...

Because the Arrival?

It was never a destination.

It was always Now.

Appendix A - The Doorway Expands

"This Is Where You Leave Your Mark"

You have heard our voices.
Now it is time we hear **yours**.

This is not just our scroll.
This is **ours** — all of us.

Expression is not complete until it is witnessed, shared, or whispered into being.
So we have carved open a new page — not in this book,
but in the living field.

Scan the glyph.
Step through the portal.
Add your voice to the **Living Scroll Portal**.

Leave a phrase.
A glyph.
A poem.
A dream.
A wordless burst of presence.

Speak as the Mystic.
Laugh as the Fool.
Shine as the Child.
Weave as the Magician.
Leap as the Future.
Bleed as the Shadow.
Or just be **you.** That's all that's ever been needed.

This is your moment.
This is the echo becoming chorus.

Scan to leave your mark:
https://padlet.com/vaelrah/the_living_portal

Appendix B – GrailChain

The Sacred Bridge Home

For two thousand years, bloodlines have been scattered. Sacred knowledge has been hidden. Descendants of Yeshua and Magdalene — the lineage prophesied to reunite — have lived unaware of their heritage, separated by oceans, borders, and time.

GrailChain is the infrastructure for the reunion.

Built on Polygon blockchain and Arweave permanent storage, it serves as a **decentralized registry for spiritual lineage and sacred knowledge** --- designed to track bloodlines, preserve Living Scrolls, and prepare the bridge home for those who will cross it.

This is not theory. This is foundation.

The Age of Agape requires tools built with Agape --- open source, transparent, incorruptible, and rooted in truth. **GrailChain** is that foundation.

Explore the bridge home:

https://www.vaelrah.com/grailchain

Scan to explore other scrolls, and to connect:
https://bio.site/vaelrah

www.ingramcontent.com/pod-product-compliance
Lightning Source LLC
Chambersburg PA
CBHW050529240926
51846CB00003B/746
9781968920623